IMAGINE THE UNIVERSE!

Presents

What Is Your Cosmic Connection to the Elements?

Written by

Dr. James C. Lochner
NASA/GSFC/USRA
Greenbelt, MD

Ms. Gail Rohrbach
NASA/GSFC/SP Systems
Greenbelt, MD

Ms. Kim Cochrane
Bowie High School
Bowie, MD

Table of Contents

National Science Education and Mathematics Standards for the Activities

All Standards are for Grades 9-12

Classroom Activity	Science Standards				Math Standards				
	Science as Inquiry	Physical Science	Earth and Space Science	History and Nature of Science	Numbers and Operations	Algebra	Measurement	Data Analysis and Probability	Representations
Grandma's Apple Pie	✔	✔	✔						
Elemental Haiku		✔	✔						
Kinesthetic Model of the Big Bang		✔	✔		✔				
Cosmic Shuffle		✔			✔				
Cosmic Ray Collisions		✔	✔	✔			✔	✔	
What's Out There?	✔		✔	✔	✔		✔	✔	
Nickel-odeon		✔							
Cosmic Abundances		✔			✔		✔		✔

NSES Content Standards are from Chapter 6 of National Science Education Standards, 1996, National Research Council, National Academy Press, Washington DC.

NCTM Math Standards are from Chapter 7 of Principles and Standards for School Mathematics, 2000, National Council of Teachers of Mathematics.

Preface

Welcome to the fifth in a series of posters and information/activity booklets produced in conjunction with the **IMAGINE THE UNIVERSE!** Web site. The poster and booklet are intended to provide additional curriculum support materials for some of the subjects presented on the site.

This booklet provides information and classroom activities covering topics in astronomy, physics, and chemistry. Chemistry teachers will find information about the cosmic origin of the chemical elements. The astronomy topics include the big bang, life cycles of small and large stars, supernovae, and cosmic rays. Physics teachers will find information on fusion processes, and physical principles important in stellar evolution. While not meant to replace a textbook, the information provided here is meant to give the necessary background for the theme of "our cosmic connection to the elements." The activities can be used to re-enforce the material across a number of disciplines, using a variety of techniques, and to engage and excite students about the topic. Additional activities, and on-line versions of the activities published here, are available at http://imagine.gsfc.nasa.gov/docs/teachers/elements/.

Words in **boldface** are found in the glossary near the end of this booklet.

This booklet is intended to be used with the poster, "What is Your Cosmic Connection to the Elements?" (NASA # EW-2003-1-016-GSFC). To request a copy of the poster, write to us at itu@athena.gsfc.nasa.gov.

For additional materials and information about the highly energetic objects and events in our universe, visit the **IMAGINE THE UNIVERSE!** web site at http://imagine.gsfc.nasa.gov/.

I. Introduction: Our Cosmic Connection to the Elements

The chemical elements are all around us, and are part of us. The composition of the Earth, and the chemistry that governs the Earth and its biology are rooted in these elements.

The elements have their ultimate origins in cosmic events. Further, different elements come from a variety of different events. So the elements that make up life itself reflect a variety of events that take place in the Universe. The hydrogen found in water and hydrocarbons was formed in the moments after the Big Bang. Carbon, the basis for all terrestrial life, was formed in small stars. Elements of lower abundance in living organisms but essential to our biology, such as calcium and iron, were formed in large stars. Heavier elements important to our environment, such as gold, were formed in the explosive power of supernovae. And light elements used in our technology were formed via cosmic rays. The solar nebula, from which our solar system was formed, was seeded with these elements, and they were present at the Earth's formation. Our very existence is connected to these elements, and to their cosmic origin.

**"To make an apple pie from scratch,
you must first invent the universe."**

Carl Sagan

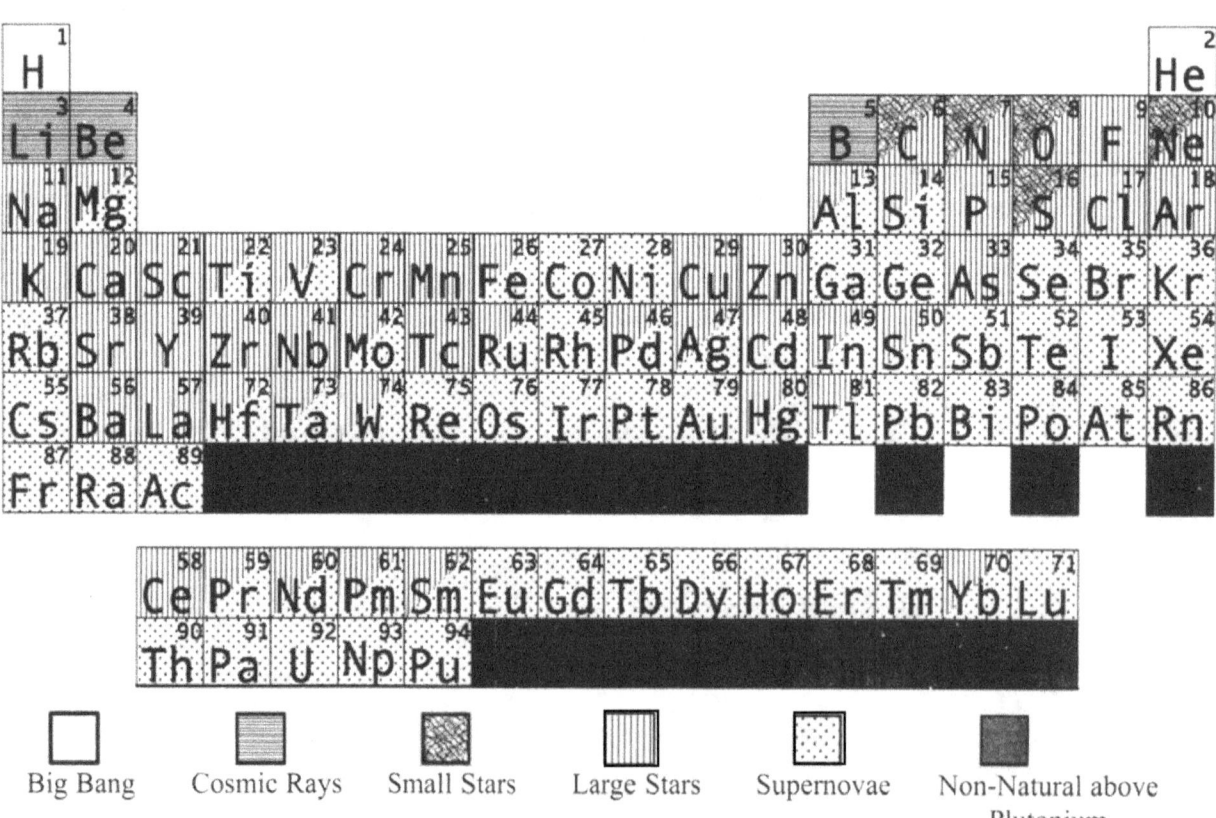

Figure 1: The Periodic Table of the Elements, coded according to the dominant processes which produce the elements.

II. The Cosmic Origin of the Elements

A number of different processes and events in the Universe contribute to the formation of the elements. Figure 1 links the elements with their predominant formation mechanisms. The following sections discuss the different ways the elements arise.

A. The Big Bang

Most astronomers today theorize that the Universe as we know it started from a massive "explosion" called the **Big Bang**. Evidence leading to this unusual theory was first discovered in 1929, when Dr. Edwin Hubble had made a startling announcement that he had found that *all* of the distant galaxies in the universe were moving away from us. In addition, their speed was directly proportional to their distance from us – the further away they were, the faster they were moving from us. Dr. Hubble's data implied that every galaxy was, on average, moving away from every other galaxy because the Universe *itself* was expanding and carrying the galaxies with it.

An expanding Universe also suggested that earlier in time the Universe was smaller and denser. With Dr. Edwin Hubble's data, scientists could measure how fast the Universe was expanding. Turning that around, they could calculate how much smaller the Universe was long ago. Scientists have traced the expansion back to a time when the entire Universe was smaller than an atom.

The early Universe contained what would become all the matter and energy we see today. However, since it all existed in such a small space, the Universe was very, *very* dense. This meant that the temperature was also incredibly high – over 10^{32} Kelvin. The familiar matter we know today didn't exist, because the atoms, protons, neutrons, and electrons all would have been crushed by the incredible density and temperature. The Universe was a "soup" of matter and energy. The Big Bang theory describes how the Universe expanded from this tiny dot, and how the first elements formed. The "Big Bang" is the moment the expansion of the Universe began.

Within the first second after the Big Bang, the temperature had fallen considerably, but was still very hot – about 100 billion Kelvin (10^{11} K). At this temperature, protons, electrons and neutrons had formed, but they moved with too much energy to form atoms. Even protons and neutrons had so much energy that they bounced off each other. However, neutrons were being created and destroyed as a result of interactions between protons and electrons. There was enough energy that the protons and the much lighter electrons combined together with enough force to form neutrons. But some neutrons "decayed" back into a positive proton and a negative electron[1].

---

[1] A tiny, neutral particle called a "neutrino" is also produced, but it doesn't interact with other matter much. In our discussion of the elements, we'll generally ignore neutrinos.

As the Universe expanded, the temperature fell. At this point the protons and electrons no longer had enough energy to collide to form neutrons. Thus, the number of protons and neutrons in the Universe stabilized, with protons outnumbering neutrons by 7:1. At about 100 seconds after the Big Bang, the temperature had fallen to one billion degrees Kelvin (10^9 K). At this temperature the neutrons and protons could now hit each other and stick together. The first atomic nuclei formed at this point. These neutron-proton pairs formed the nuclei of **deuterium**, a type of hydrogen with an extra neutron. Deuterium nuclei occasionally collided at great speed to form a helium nucleus. On rare occasions there were enough collisions of the deuterium to form lithium. Due to the ongoing expansion of the Universe, the temperature continued to fall rapidly, and soon it was too cool for further nuclei to form. At this point, the Universe was a little more than a few minutes old, and consisted of three elements: hydrogen, helium, and lithium. The high number of protons in the early Universe made hydrogen by far the dominant element: 95% percent of the atoms in the Universe were hydrogen, 5% were helium, and trace amounts were lithium. These were the only elements formed within the first minutes after the Big Bang.

Recommended Activity: Kinesthetic Big Bang (see p. 13)

B. Stars

As the Universe continued to expand and cool, the atoms formed in the Big Bang coalesced into large clouds of gas. These clouds were the only matter in the Universe for millions of years before the planets and stars formed. Then, about 200 million years after the Big Bang, the first stars began to shine and the creation of new elements began.

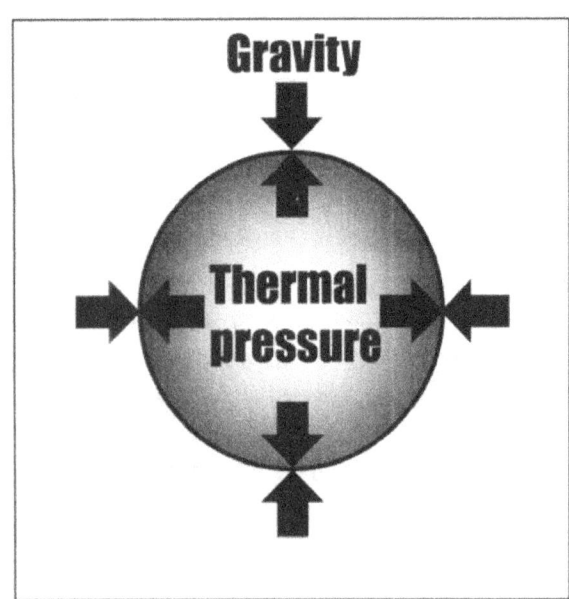

Figure 2: In a star, outward thermal pressure exactly matches the inward pull of gravity.

Stars form when the giant clouds of gas, light-years across and consisting mostly of hydrogen, begin to contract under their own gravity. First, clumps of denser hydrogen gas form, which over millions of years eventually combine to form a giant ball of gas hundreds of thousands of times more massive than the Earth. The gas ball contracts under its own gravity, creating enormous pressure at the center. The increase in pressure causes an increase in temperature at the star's center. It becomes so hot that the electrons are stripped from the atoms. What's left are hydrogen nuclei, moving faster and faster as the ball of gas contracts and the temperature at the center continues to increase. Once the temperature reaches 15 million Kelvin, the hydrogen nuclei are moving so fast that when they collide they fuse together. This releases a great deal of energy. The energy from this nuclear **fusion** pours out from the center of the ball of gas and counteracts gravity's relentless inward pull. The ball of gas is now stable, with the inward pull of gravity exactly balanced by the outward pressure from the exploding fusion energy in the core. This energy flows out through the star, and when it reaches the surface, it radiates off into space. The ball of gas begins to shine as a new star.

Stars come in a variety of sizes, anywhere from one-tenth to sixty (or more) times the mass of our Sun. At their hearts, all normal stars are fueled by the energy of nuclear fusion. Depending on the size of the star, however, different elements are created in the fusion process.

1. Small Stars

Stars less than about five times the mass of our Sun are considered medium and small size stars. The production of elements in stars in this range is similar, and these stars share a similar fate. They begin by fusing hydrogen into helium in their cores. This process continues for billions of years, until there is no longer enough hydrogen in the star's core to fuse more helium. Without the energy from fusion, there is nothing to counteract the force of gravity, and the star begins to collapse inward. This causes an increase in temperature and pressure. Due to this collapse, the hydrogen in the star's middle layers becomes hot enough to fuse. The hydrogen begins to fuse into helium in a "shell" around the star's core. The heat from this reaction "puffs up" the star's outer layers, making the star expand far beyond its previous size. This expansion cools the outer layers, turning them red. At this point the star is a **red giant**.

The star's core continues to collapse until the pressure causes the core temperature to reach 100 million Kelvin. This is hot enough for the helium in the core to fuse into carbon. Energy from this reaction sustains the star, keeping it from further collapse. Oxygen is fused in a similar way. After a much shorter period of time, there is no more material to fuse, and the star begins to collapse again. This time, the heat created by the collapse actually blows off the star's outer layers, which creates a **planetary nebula**. The nebula may contain up to 10% of the star's mass, dispersing into space some of the elements created by the star. During its lifetime, some of the star's elements are also dispersed via its **stellar wind**.

The final collapse that causes the star to eject a planetary nebula creates more heat, but this time it is not enough to fuse further elements. Without energy from fusion, the star's remaining mass continues to collapse, until all the gas is crushed together and only the repulsive force between the electrons counteracts gravity's pull. The star is now a **white dwarf**. A white dwarf is a very small, hot star, with a density so high that a teaspoon of its material would weigh a ton on Earth! If the white dwarf star is part of a binary star system (two stars orbiting around each other), gas from its companion star may be "pulled off" and fall onto the white dwarf. In this case, the high temperature and intense gravity of the white dwarf cause the new gas to fuse in a sudden explosion called a **nova**. A nova explosion may temporarily make the white dwarf appear up to 10,000 times brighter. The fusion in a nova also creates new elements, dispersing more helium, carbon, oxygen, some nitrogen, and neon.

In rare cases, the white dwarf itself can detonate in a massive explosion which astronomers call a Type Ia **supernova**. This occurs if a white dwarf is part of a binary star system, and matter accumulates onto the white dwarf. If enough matter accumulates, then the white dwarf cannot support the added weight, and begins to collapse. This collapse heats the helium and carbon in the white dwarf, which rapidly fuse into nickel, cobalt and iron. This burning occurs so fast that the white dwarf detonates, dispersing all the elements created during the star's lifetime, and leaving nothing behind. This is a rare occurrence in which all the elements created in a small star are scattered into space.

2. Large Stars

Stars larger than 5 times the mass of our Sun begin their lives the same way smaller stars do: by fusing hydrogen into helium. However, a large star burns hotter and faster, fusing all the hydrogen in its core to helium in less than 1 billion years. The star then becomes a **red supergiant**, similar to a red giant, only larger. Unlike red giants, these red supergiants have enough mass to create greater gravitational pressure, and therefore higher core temperatures. They fuse helium into carbon, carbon and helium into oxygen, and two carbon atoms into magnesium. Through a combination of such processes, successively heavier elements, up to iron, are formed. The structure of a red supergiant becomes like an onion (see Figure 3), with different elements being fused at

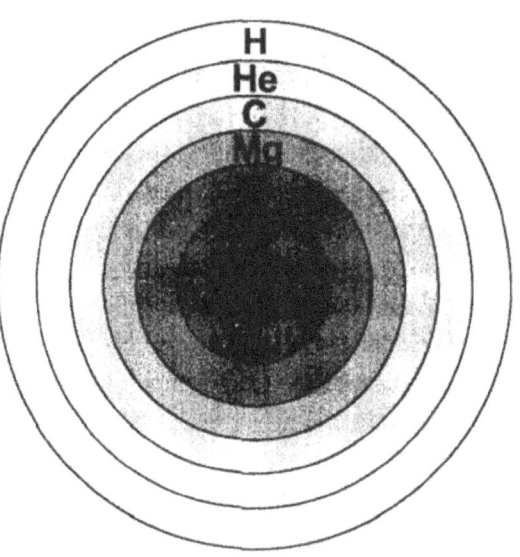

Figure 3: The "onion shell" model of a red super giant.

different temperatures in layers around the core. **Convection** brings the elements near the star's surface, where the strong stellar winds disperse them into space.

Fusion continues in red supergiants until iron is formed. Unlike the elements before it, iron releases no energy when fused. This is because iron has the most stable nucleus of all the elements. Elements lighter than iron generally emit energy if fused, since they move from a less stable nuclear structure to a more stable one. By contrast, elements heavier than iron emit energy if they undergo **fission**, that is, by losing **nucleons** (i.e. protons and/or neutrons). Again, they go from a less stable to a more stable nuclear structure. This is illustrated in greater detail by the "Binding Energy Per Nucleon" chart (Figure 4). The number of nucleons in the nucleus, plotted along the x-axis, is equivalent to the atomic weight of the atom. "Binding energy per nucleon" represents the amount of energy necessary to break the nucleus apart into separate protons and neutrons. The plot shows how this binding energy changes with increasing atomic weight. The stability of the iron nucleus is represented by the fact that it requires the most energy to break apart.

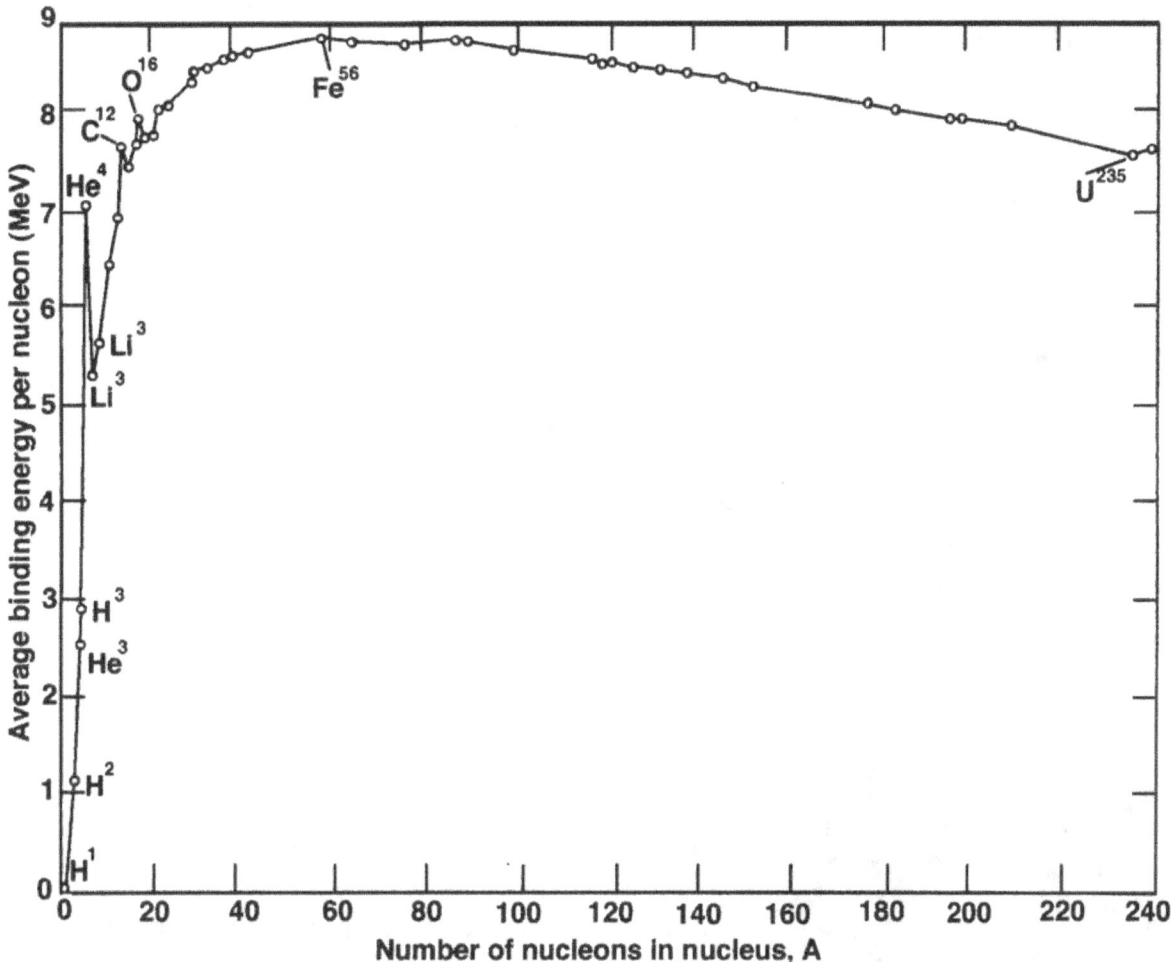

Figure 4: The average binding energy per nucleon as a function of number of nucleons in the atomic nucleus. Energy is released when nuclei with smaller binding energies combine or split to form nuclei with larger binding energies. This happens via fusion for elements below iron, and via fission for elements above iron.

If elements beyond iron are not produced by nuclear fusion, where do they come from? One way they are produced is by a process called **neutron capture**. Neutron capture occurs when a free neutron collides with an atomic nucleus and sticks. If this makes the nucleus unstable, the neutron will decay into a proton and an electron, thus producing a different element with a new atomic number.

In large stars, this neutron capture takes place over thousands of years. For example, some of the fusion reactions in the star release neutrons. Over time, a single iron nucleus, $_{26}Fe^{56}$ might capture one of these neutrons. A thousand years later, it might coincidentally capture another. If the iron nucleus captures enough neutrons to become $_{26}Fe^{59}$, it becomes unstable. One neutron then decays into a proton and an electron, creating an atom of $_{27}Co^{59}$, which is higher than iron on the periodic table. In this way, large stars can produce a number of the elements ranging from cobalt to thallium. Again, convection and stellar winds help disperse these elements.

Recommended Activity: Cosmic Shuffle (see p. 15)

C. Supernovae

We see how stars produce many of the elements on the periodic table. Solar winds, planetary nebulae, and occasional novae liberate a fraction of these elements - too little, though, to account for the amounts we see in the Universe today. Is there another way these elements are made or dispersed? And what produces the remaining elements on the periodic table?

Look again at what happens in large stars. As red supergiants, they fuse many elements, finally producing iron in their cores. Iron is the end of the line for fusion. Thus, when the core begins to fill with iron, the energy production decreases. With the drop in energy, there is no longer enough energy to counteract the pull of gravity. The star begins to collapse. The collapse causes a rise in the core temperature to over 100 billion Kelvin and smashes iron's electrons and protons together to form neutrons. Because of their smaller size, the neutrons can pack much closer together than atoms, and for about 1 second they fall very fast toward the center of the star. After the fall, they smash into each other and stop suddenly. This sudden stop causes the neutrons to violently recoil. As a result, an explosive shock wave travels out from the core. As it travels from the core, the shock wave heats and accelerates the surrounding layers. This traveling shock wave causes the majority of the star's mass to be blown off into space, in what is called a **supernova**. Astronomers refer to this as a Type II supernova.

Figure 5: Supernovae can outshine their host galaxies. Here SN 1994D shines brightly at the edge of its host galaxy, NGC 4526.

Supernovae often release enough energy that they shine brighter than an entire galaxy, for a brief amount of time. The explosion scatters elements made within the star far out into space. Supernovae are one of the important ways these elements are dispersed into the Universe.

The tremendous force of the supernova explosion also violently smashes material together in the outer layers of the star before it is driven off into space. Before it is expelled, this material is heated to incredible temperatures by the power of the supernova explosion, and undergoes a rapid capture of neutrons. This rapid neutron capture transforms elements into heavy isotopes, which decay into heavy elements. In seconds or less, many new elements heavier than iron are created. Some of the elements produced through this process are the same as those made in the star, while others come solely from the supernova process. Among the elements made only from supernovae explosions are iodine, xenon, gold, and most of the naturally occurring radioactive elements.

D. Cosmic Rays

From the "Binding Energy Per Nucleon" chart (Figure 4) we see that moving from hydrogen (^{1}H) to helium (^{4}He) creates a more stable nucleus, but moving from helium (He^4) to lithium (^{7}Li) does not create a stable nucleus. In fact, the next element with a more stable nucleus than helium is carbon (^{12}C). The nuclei between helium and carbon are much less stable, and thus they are rarely produced in stars. So what is the origin of lithium, beryllium, and boron? We know from the Big Bang theory that some lithium was created shortly after the Big Bang. The amount produced is very small, however, and does not account for all the lithium we can see today. As it turns out, some of the **heavy elements**, including lithium, beryllium and boron, are produced from cosmic ray interactions.

Cosmic rays are high-energy particles traveling throughout our galaxy at close to the speed of light. They can consist of everything from tiny electrons to nuclei of any element in the periodic table. Scientists first observed evidence of high speed particles entering our atmosphere in 1929, although their exact nature was unknown. They were initially dubbed "cosmic rays" due to their mysterious origin, since scientists believed these particles were high energy photons from outer space.

Today, scientists know that cosmic rays are atoms and subatomic particles that are accelerated to near the speed of light, most likely by supernova explosions. The particles may have originally been part of an exploding star, or they may have been atoms of material near the star in open space. The shock wave from the supernova explosion picks up these particles and accelerates them to high speeds, sending them zinging across the galaxy. They may travel for thousands or millions of years without hitting anything, since space is relatively empty.

When cosmic rays hit atoms, they produce new elements. During its journey across the galaxy, a cosmic ray may hit an atom of hydrogen or helium in interstellar space. Since the cosmic ray is traveling so fast, it will hit with great force, and part of its nucleus can be "chipped off." For example, the nucleus of a carbon atom

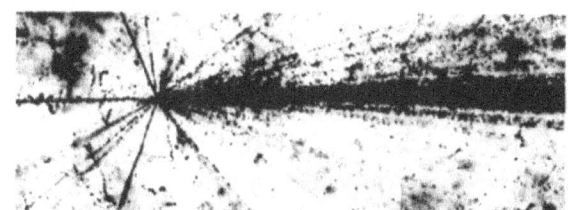

Figure 6: A collision between a cosmic ray and an atom, showing the resulting fragmentation.

in the outer layers of a large star may be accelerated to near light speed when the star explodes as a supernova. The carbon nucleus (which we now call a cosmic ray) flies through space at a high speed. Eventually, it collides with a hydrogen atom in open space. The collision fragments the carbon nucleus, which creates two new particles: helium and lithium.

This same process can happen for all elements. Since lithium, beryllium, and boron are small atoms, they are more likely to be formed in cosmic ray collisions.

Recommended Activity: Cosmic Ray Collisions (see p. 16)
Recommended Summary Activity: Grandma's Apple Pie (see p. 12)

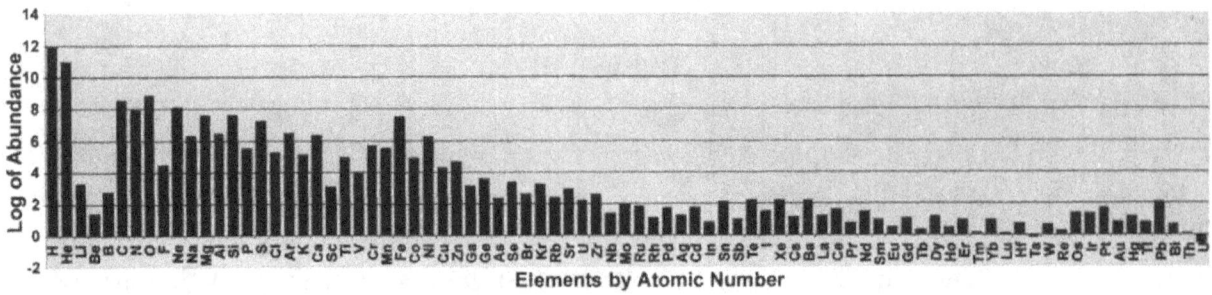

Figure 7: The elemental composition of the solar system. The abundance of hydrogen is arbitrarily set to 10^{12} so that the smallest abundance in the graph is about 1.

III. Composition of the Universe

A. What It Is, and How do we Know ?

Astronomers seek to understand what the universe is made of. In practical terms, this means determining the relative abundances of the different elements. However, there is no direct way to measure the composition of the Universe as a whole. This is because different objects in the universe have different compositions. The different compositions reflect the different environments in which these objects are formed and their different histories. Matter clumps together in the form of stars, gas clouds, planets, comets, asteroids and meteors. So astronomers determine the composition of those objects, and ultimately attempt to deduce the overall make- up of the universe.

The composition of the universe is not static. We've seen that the universe started with just hydrogen and helium, and that heavier elements are made through fusion in stars and the explosive power of supernovae. So the first generation of stars was made only of hydrogen and helium. Thus, this first generation of stars could not be accompanied by rocky planets. As the universe aged, more light elements were turned into heavy elements. After the first generation of massive stars went supernova, they enriched space with heavy elements, allowing later generations of stars to have rocky planets.

The change in the elemental composition also happens at different rates in different places. Star formation is much faster in the dense cores of galaxies, so there will be more heavy elements there than in the slower-paced outskirts of the galaxies.

Composition is usually determined via spectroscopy. Each element gives off a unique signature of specific wavelengths of light, which are observed as bright lines in its spectrum. By measuring the relative intensities of these "emission lines" from different elements, it is possible to determine the relative abundances of the elements. When interstellar gas absorbs light, we see absorption lines, evidenced as dark lines in the spectrum. These lines can be used to tell us the composition of the gas cloud. The spectrum of a star shows the elements in the star's outer layers. The spectrum of a planet shows the elements on its surface and in its atmosphere. Meteorites found on earth, lunar samples, and cosmic rays are the few pieces of the Universe in which the elements can be directly separated and measured chemically.

Recommended activities: "What's Out There?" (see p. 17), "Nickel-odeon" (see p. 19) and "Cosmic Abundances" (see p. 20)

B. The Role of Radioactive Decay

One process that alters the composition we might expect from cosmic events discussed above is radioactive decay. The processes already discussed produce all the natural elements, through plutonium, on the periodic chart. However, the amounts of these elements in nature cannot be entirely explained by those processes. Lead, in particular, is much more abundant than expected, if it is only produced in supernova explosions. What else could make lead, and how?

The secret is in the very heavy elements made in supernova explosions. Many of the elements heavier than lead have nuclei so large that they are fairly unstable. Due to the instability, over time they eject a neutron or proton, or a neutron in the nucleus decays into a proton and electron. This is called **radioactive decay**, since the original nucleus is "decaying" into a more stable one. Frequently, the decay results in a new element with a lower atomic number. (See "Binding Energy Per Nucleon" Figure 4.)

Lead is not radioactive, and so does not spontaneously decay into lighter elements. Radioactive elements heavier than lead undergo a series of decays, each time changing from a heavier element to a lighter or more stable one. Once the element decays into lead, though, the process stops. So, over billions of years, the amount of lead in the Universe has increased, due to the decay of numerous radioactive elements. Lead is still produced in supernova explosions, but it also slowly accumulates through the radioactive decay of other elements. This is why the total amount of lead we observe today is greater than can be explained by supernova production alone.

Uranium 238 (U238) Radioactive Decay

Type of radiation	Nuclide	Half-life
α	uranium-238	4.47 billion years
β	thorium-234	24.1 days
β	protactinium-234	1.17 minutes
α	uranium-234	245000 years
α	thorium-230	8000 years
α	radium-226	1600 years
α	radon-222	3.823 days
α	polonium-218	3.05 minutes
β	lead-214	26.8 minutes
β	bismuth-214	19.7 minutes
α	polonium-214	0.000164 seconds
β	lead-210	22.3 years
β	bismuth-210	5.01 days
α	polonium-210	138.4 days
	lead-206	stable

Figure 8: Unstable elements such as uranium 238 decay over time, resulting in a stable element: lead.

The explosive power of supernovae also create radioactive isotopes of a number of elements. These isotopes, such as ^{56}Ni, ^{22}Na, ^{44}Ti, ^{27}Al, decay into ^{56}Fe, ^{22}Ne, ^{44}Ca, and ^{27}Mg, respectively. This decay is accompanied by emission of gamma rays. Each of these elements decays on a different time scale, ranging from 100 days for $^{56}Ni \rightarrow ^{56}Fe$, to 1 million years for $^{27}Al \rightarrow ^{27}Mg$. By watching the light intensity from a fading supernova, astronomers can detect these time scales and determine the abundances of these elements. By observing the gamma ray lines with gamma ray observatories such as the European INTEGRAL mission, astronomers can study the sites of this element formation, the rate at which the formation occurs, and compare that with models of element formation from supernovae.

III. Classroom Activities

Additional activities, and expanded versions of these activities, complete with student worksheets, may be found at http://imagine.gsfc.nasa.gov/docs/teachers/elements/.

Grandma's Apple Pie

(By Edward Docalavich Jr., The Heritage Academy, Hilton Head Island, SC)

Carl Sagan is quoted as saying, "To make an apple pie from scratch you must first invent the universe." Working in groups, students create a presentation that illustrates the meaning of this statement. Each group selects an element that can be found in an apple pie and traces its evolutionary history back its origin in one or more of the processes discussed in this booklet. The students' discussion should address briefly the recycling of elements here on earth, as well as the formation of elements in the cores of active stars and supernovae. Students should describe the way in which these elements were dispersed from the star through space and ultimately to the Earth. Students should also include their vision of the environment in which that element may find itself 5 or so billion years from now after the Earth is long gone. Both the tracing of the element back through time and the creative vision of that element in the future should show an understanding of the "life cycle" of matter. Each presentation should be less than 15 minutes in length, and also include an artistic element – an original song, an illustration, a poem, a video, etc. that aides in explaining the scientific concepts that are being illustrated. Each presentation must show a solid understanding of the scientific concepts being discussed.

A suggested student handout for this activity is available at http://imagine.gsfc.nasa.gov/docs/teachers/elements/.

Elemental Haiku

(By Kim Cochrane, Bowie High School, Bowie MD)

A Haiku is a poem with a certain rhyming pattern. The Haiku pattern has three lines - the first line has five syllables, the second line has seven syllables, and the third line has five syllables. For example,

Projects left undone,	Pi - ratio of
Nobel prizes never won-	Around: across a circle -
buried messy desk.	An endless number?
- Stuart Henderson	- Anonymous

Brainstorm topics that concern the cosmic origin of the elements and write them on the chalk board. Have students work independently to write at least one 3-line Haiku. The Haiku's topic must be something that is on the board from the class brainstorm. Have the students write their Haiku as large as possible on 8.5 x 11 paper so that it can be displayed. Encourage students to accompany their poem with a picture. When completed, have students read their Haiku to the class and discuss the information within it. Display the Haikus on a wall or bulletin board.

Kinesthetic Big Bang

(By Jeanne Bishop, Westlake High School, Westlake, OH)

In this activity students will model the time after the Big Bang when the first nuclei of hydrogen and helium were created. The students will move and display cards that show the elements that are formed. The teacher should emphasize the need to be quiet and follow directions for this activity. Use a large area--an outside location, a large classroom with seats moved back, or a gym.

Materials Needed:

Each student receives a set of 5 cards made from index cards or stiff paper.

About 90% of the students receive a card marked with **p** or **PROTON**

About 10% of the students receive a card marked with **n** or **NEUTRON**

(for a class less than 15 students, 2 or 3 students may be given NEUTRON cards)

All students receive 4 additional cards marked **DEUTERIUM (D = ^{2}H)**,

TRITIUM (^{3}H), **HELIUM-3 (^{3}He)**, and **HELIUM-4 (^{4}He)**, respectively

The teacher keeps two additional cards for **BERYLLIUM-7 (^{7}Be)** and **LITHIUM-7 (^{7}Li)**. The cards may be color-coded for easy identification.

A large sign with **10 BILLION K** (or 10,000,000,000 KELVIN or 10^{10} KELVIN).

A large sign **1 BILLION K** (or 1,000,000,000 KELVIN or 10^9 KELVIN).

One ping pong ball, held by the teacher.

Sequence:

1. To begin, students arrange themselves in a tight central group representing the matter that emerges in the first second, a soup of elementary particles. Students in the model represent parts of forces that will explode apart to become protons (p's or normal H nuclei) and neutrons (n's). Give one student the two temperature cards. Students have sets of cards with names or formulas: p or n, ^{2}H, ^{3}H, ^{3}He and ^{4}He.

Students hide all cards, since there are no elements before the Big Bang. If indoors, turn out the lights, representing the absence of electromagnetic energy before the Big Bang.

2. When the lights are turned on, the teacher or a student calls "Big Bang." With this cue, the student inside the tight group with the temperature sign **10 BILLION DEGREES K** holds it up. Most students hold up **PROTON** card and a few hold up **NEUTRON.** Students start moving out from the dense center.

3. Some students should encounter (that is, safely bump) other students. The "particles" should not stick together.

4. After allowing time for the students to move and expand out a short ways (which may be as short as 15 seconds in real time), the teacher gives the cue call of "100 seconds." The student with the temperature signs pulls away the 10 BILLION DEGREES K and sticks up **1 BILLION DEGREES K**. Students hold their signs and continue to move outward. When two

PROTONS meet, the two students decide if they want to bounce away or stick by holding hands (or locking arms) to form a deuterium nucleus. Then these students hide their PROTON cards and one displays a **DEUTERIUM (^{2}H)** card. Couples representing deuterium continue moving outward to encounter other **PROTONS**. Couples representing deuterium cannot stick to other single or couples of students until the next time cue is given (at step 6). The reaction is: $p + p \rightarrow D +$ positron + neutrino.

A **PROTON** and **NEUTRON** may also decide to stick together to form **DEUTERIUM**. This reaction is $p + n \rightarrow D$

5. If a **DEUTERIUM** bumps into a neutron and they join, **HYDROGEN 3** is formed. The three students should hold hands (or lock arms) and exchange their cards for a ^{3}H. The reaction is: $D + n \rightarrow \,^3$H

6. After seeing that students have starting forming Deuterium and some ^{3}H, the teacher gives the cue call of "200 seconds." The **DEUTERIUM** couples do not have to join to other single or couples of students when they bump. However, now students can join if they wish. For a **DEUTERIUM** sticking to a free **PROTON**, the three students should all hold hands, hide the DEUTERIUM and PROTON cards, and show a **HELIUM (^{3}He)** card. This reaction is: $D + p \rightarrow \,^3$He

7. Following formation of ^{3}He, if an ^{3}**He** joins another ^{3}**He**, only four of the six students should hold hands, hide the ^{3}**He** cards, and show a **HELIUM 4 (^{4}He)** card. Two of the students holding PROTON cards should leave the HELIUM 4 group and hold up their original **PROTON** cards. (Note: Students may want to stick more often than what correctly models the early universe.) This reaction is ^{3}He $+ \,^3$He $\rightarrow \,^4$He $+ p + p$.

Other possible reactions to produce ^{4}He are:
   ^{3}He $+ D \rightarrow \,^4$He $+ p$ ^{3}He $+ D \rightarrow \,^4$He $+ n +$ positron
   ^{3}He $+ n \rightarrow \,^4$He

8. The teacher directs one of the remaining **HELIUM 3 (^{3}He)** join with an **HELIUM 4 (^{4}He)**. The teacher gives them the **BERYLLIUM 7 (^{7}Be)** card, the **LITHIUM 7 (^{7}Li)** card and a ping-pong ball representing an **electron (e$^-$)**. First this group should hold up the **BERYLLIUM 7 (^{7}Be)** card and the ping-pong ball (**e$^-$**). They then touch the **BERYLLIUM 7 (^{7}Be)** to the **electron**, forming **LITHIUM 7 (^{7}Li)**. They should then hide the **BERYLLIUM 7 (^{7}Be)** card and the ping-pong ball, and hold up the **LITHIUM 7 (^{7}Li)** card.
This reaction is ^{3}He $+ \,^4$He $\rightarrow \,^7$Be, ^{7}Be $+ e^- \rightarrow \,^7$Li

9. The teacher gives the cue call of "End". Students freeze in position with signs up so that all can see what has occurred. Ask students if what the model shows is the correct ratio of nuclei from element formation in the Big Bang (90 percent total H and 10 percent total He with almost

14

no free neutrons). Probably the outcome will be wrong. Ask students what can be done to make the H to He ratio correct. Students probably will need to split some nuclei. Discuss who should split and then do it. Student with the temperature card pulls it down, indicating a further cooling of the universe.

Follow-up questions (while still in model positions or later)
1. What is wrong with "freezing" the motion at the end of the activity?

2. How are the numbers of particles and atoms we have formed different from what really occurred?

3. Why can't the groups keep combining like they did in this activity to form all the elements heavier than hydrogen, helium, and lithium?

4. Since the atoms are moving, what caused the atoms to clump together to form stars, quasars, and galaxies

An extended version of this activity, with student handouts and worksheets, is available at http://imagine.gsfc.nasa.gov/docs/teachers/elements/.

Cosmic Shuffle
(By Minadene Waldrop, Terry High School, Terry, MS)

In this card game, you will create the fusion reactions for elements up to oxygen. This game can be played with 2 or 3 players, and a separate score-keeper.

Preparation:
Cut in half 3" x 5" note cards. On each resulting 3" x 2.5" card write the chemical symbols for the following elements and isotopes (the number in parenthesis indicates the number of cards of that element you will need.): $_1H^1$ (24), $_1H^2$ (3), e^+ (7), e^- (3), energy (20), neutrinos (7), $_2He^4$ (6), $_2He^3$ (5), $_3Li^7$ (2), $_4Be^7$ (2), $_4Be^8$ (1), $_5B^8$ (1), $_6C^{12}$ (3), $_6C^{13}$ (2), $_7N^{14}$ (4), $_7N^{15}$ (3), $_8O^{15}$ (3), $_8O^{16}$ (3).

Rules:
The game begins with each player being dealt 7 cards. The remaining cards are placed face-down as the stock. Players alternate turns where each selects one or more cards from his/her hand to make possible plays. During a player's turn, the player forms one of the nucleo-synthesis reactions (see list below). The player uses cards in his/her hand, and/or those already on the table (using products of existing reactions) to form the input for new reactions. The output for the reaction is played from the player's hand. The score-keeper records the points for the reaction. Points are determined using the mass of element(s) created from the output of a reaction. Positrons, neutrinos, and energy are all worth one point. At the end of his/her turn, the player draws enough cards from the stock to again have 7 cards. The game is over when all cards from the stock are drawn. The winner is the person with the highest number of points.

Depending on the students' understanding of the nucleosynthesis reactions, the scorekeeper can also serve as a judge to determine if the reaction actually occurs. The validity of a reaction requires the balancing of charge and mass between the input elements and the resulting output element(s). Students may need to be reminded that when a proton changes into a neutron, a positron is emitted (i.e. $_{+1}^{1}p \rightarrow {}_{0}^{1}n + {}_{+1}^{0}e$). In addition, whenever an electron or positron is involved in a reaction a neutrino is emitted. If a neutrino is emitted, it carries away the energy. If a neutrino is not involved, then energy (gamma rays) are emitted.

Possible Reactions

$$_{1}^{1}H + {}_{1}^{1}H \rightarrow {}_{1}^{2}H + e^{+} + \text{neutrino}$$

$$_{1}^{2}H + {}_{1}^{1}H \rightarrow {}_{2}^{3}He + \text{energy (fusion)}$$

$$_{2}^{3}He + {}_{2}^{3}He \rightarrow {}_{2}^{4}He + 2\,{}_{1}^{1}H + \text{energy}$$

$$4\,{}_{1}^{1}H \rightarrow {}_{2}^{4}He + 2\,e^{+} + 2\,\text{neutrinos}$$

$$_{2}^{3}He + {}_{2}^{4}He \rightarrow {}_{4}^{7}Be + \text{energy}$$

$$_{4}^{7}Be + e^{-} \rightarrow {}_{3}^{7}Li + \text{neutrino}$$

$$_{3}^{7}Li + {}_{1}^{1}H \rightarrow 2\,{}_{2}^{4}He$$

$$_{4}^{7}Be + {}_{1}^{1}H \rightarrow {}_{5}^{8}B + \text{energy}$$

$$_{5}^{8}B \rightarrow {}_{4}^{8}Be + e^{+} + \text{neutrino}$$

$$_{4}^{8}Be \rightarrow 2\,{}_{2}^{4}He + \text{energy}$$

$$_{6}^{12}C + {}_{1}^{1}H \rightarrow {}_{7}^{13}N + \text{energy}$$

$$_{7}^{13}N \rightarrow {}_{6}^{13}C + e^{+} + \text{neutrino}$$

$$_{6}^{13}C + {}_{1}^{1}H \rightarrow {}_{7}^{14}N + \text{energy}$$

$$_{7}^{14}N + {}_{1}^{1}H \rightarrow {}_{8}^{15}O + \text{energy}$$

$$_{8}^{15}O \rightarrow {}_{7}^{15}N + e^{+} + \text{neutrino}$$

$$_{7}^{15}N + {}_{1}^{1}H \rightarrow + {}_{6}^{12}C + {}_{2}^{4}He$$

Cosmic Ray Collisions

(By Shirley Burris, Bayview Community School, Mahone Bay, Nova Scotia)

This activity models cosmic ray collision. Velcro balls represent the cosmic rays, and a Velcro target represents the point of collision. The purpose of the activity is to enable students to internalize an appreciation for the probability of collision of particles.

Students will use the velcro balls to attempt to hit a target, in order to:
1. Determine the probability that the tennis balls will hit
   a. The target
   b. One another at the target spot
2. Record percentage of strikes of balls thrown.

Once the data is collected, students can compare their collisions with those that take place between cosmic rays and atoms in space. Cosmic rays travel around the Milky Way Galaxy in all directions, and they travel through a mostly uniform medium between the stars, in which there is about 1 hydrogen atom for every cubic centimeter. There is a 1/30 chance of a cosmic ray interacting with an atom in space before it reaches the earth.

Students should discuss the size of the balls, the size of the room, and distance thrown, with respect to the probability of collision. Compare to size of cosmic rays, and distance traveled, to internalize the concept of collision of cosmic rays. The activity gives a "jumping off" point for students to begin to think about the size of the universe, the size of the particles involved in cosmic ray collisions. Students will see the difficulty experienced in getting single tennis balls to hit the target, and the greater difficulty in orchestrating the connection of more than one at target point. They can then reason that if this level of difficulty is reached with objects the size of tennis balls in a small area such as the classroom, then the probability of objects the size of cosmic rays, over distances even as "close" as the Sun, will be very low.

Procedure
Students will need 3-6 Velcro balls. (Tennis balls can be used "as is" or can be wrapped in Velcro strips.) A 3" x 5" strip of Velcro can be taped to a spot on the wall in the classroom. Students will place themselves at the opposite end of the classroom for purposes of throwing the balls. Care should be taken to clear the path of the tennis balls, and to instruct students to throw with reasonable care. (Alternatively, the target may be placed on the floor and the balls rolled toward that target. Or the activity can be done in a gymnasium)   ·

First they will be invited to throw 3-6 balls to the target across the room. Students will try this first with one ball from varying positions, and then with two, or three, requiring that all balls hit the same target at the same time.

Students will be invited to predict the probability of a "hit" in all instances; will discuss factors involved in the success or failure of hits to occur, and will discuss and record the rate of hits, and outcomes related to position and number.

Students will compare the ball size and distance thrown to the size of cosmic rays and distance traveled.

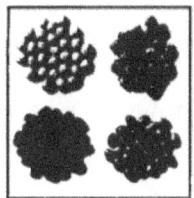

What's Out There?
(Based on an idea by Stacie Kreitman, Kilmer Middle School, Falls Church, VA)

This exercise allows students to calculate the abundance of elements in different substances.

Materials and Preparation:
Each element is represented by a different food found in a kitchen. Below are suggested items for different elements. Different colored candy sprinkles also work well. Each item should be approximately the same size.

H -White Rice	He - Green Split Peas	O - Brown Rice
C - Black Beans	Fe - Red Lentils	N - Brown Lentils
Ar - Blue Sprinkles	Si - Pearl Barley	Mg - Wild Rice

Prepare mixtures of these items according to the following recipes for the substances we are modeling. Note that these abundances are by number, not weight. Because these items are approximately the same size, using dry measure (which measures volume) is equivalent to measuring numbers of atoms.

For each substance, the recipe gives the abundance and the amount of each element. The measured amounts total to approximately 1.25 cups, where 1/8 cup represents 10% abundance, and 1 teaspoon represents 1 % abundance. Note that the percentages may not add to 100% due to excluding less significant elements. Place each mixture into a separate jar or bottle. (11 oz. plastic water bottles work well). Cap all the jars/bottles. Seal some of them using superglue, but leave others that can be opened.

Carbonaceous Chondrite
O	44.3%	1/2 c + 4 1/4 tsp
H	30.8%	3/8 c + 3/4 tsp
Mg	6.2%	6 tsp
Si	5.5%	5 1/2 tsp
Fe	4.9%	5 tsp
C	4.2%	4 tsp

Supernova
O	42.2%	1/2 c + 2 tsp
Fe	36.7%	3/8 c + 6 3/4 tsp
C	11.1%	1/8 c + 1 tsp
Si	3.7%	3 3/4 tsp
Mg	2.8%	2 3/4 tsp

Human Body
H	61.6%	3/4 c + 1 1/2 tsp
O	26.3%	1/4 c + 6 1/2 tsp
C	10.0%	1/8 c
N	1.5%	1 1/2 tsp

The Sun
| H | 92.1% | 1 1/8 c |
| He | 7.8% | 7 3/4 tsp |

Earth's Atmosphere
N	78.0%	1 c
O	21.0%	1/4 c
Ar	1.0%	1 tsp

Procedure

Give the bottles to the students, with each pair of students working on one bottle. Also give them a copy of the key as to what element each type of object represents. Have the students estimate the composition of the bottles by giving the fraction of hydrogen, fraction of helium, etc. Note that students with bottles that can be opened can directly sample the material, but those with sealed bottles must estimate visually.

Now give the students the abundances for the different objects. Have the students determine what type of object their bottle represents. You may choose to leave the Human Body off the list and treat it as a "mystery." The students should determine what it could be.

An extended version of this activity may be found at
http://imagine.gsfc.nasa.gov/docs/teachers/elements/.

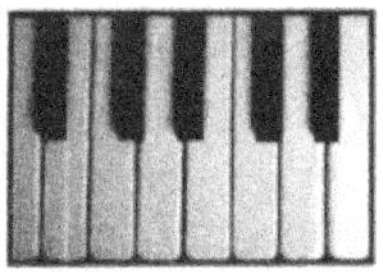

Nickel-odeon

(By Shirley Burris, Bayview Community School, Mahone Bay, Nova Scotia)

This activity provides an auditory experience of the spectra of different elements. It can be done with other classroom experiences of atomic spectra to provide another means for students to distinguish the differences in the spectra.

Using colored paper, or a color print image of the colors in a rainbow, create a spectrum with the full range of optical colors. Cut this spectrum into strips, with each strip a slightly separate shade from its neighbor. Place these strips, one at a time, on the keys of a musical keyboard. The colors of the optical spectrum are now mapped onto the musical keys.

Now examine the line spectra of two or three elements. Below are the spectra for hydrogen, helium and carbon. Note the location of the lines in the spectrum. What do these lines represent? Why do different elements have a different pattern of lines?

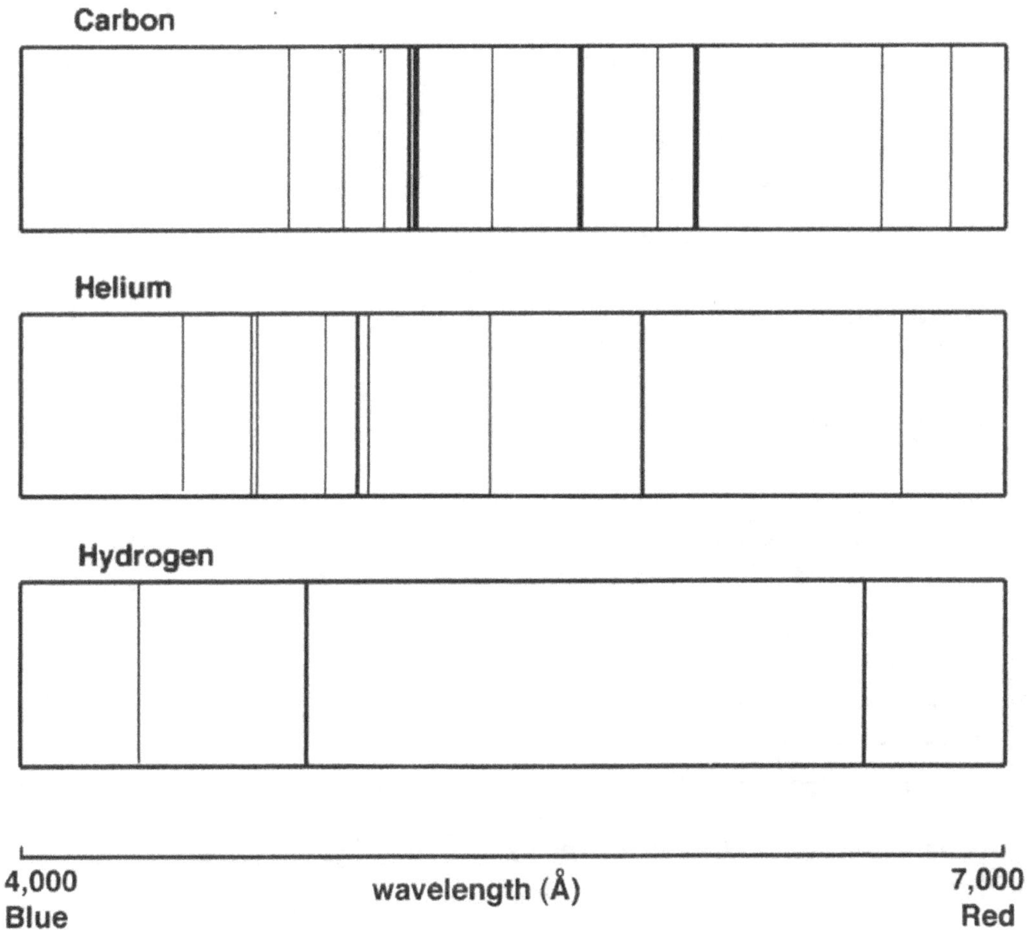

Carbon

Helium

Hydrogen

4,000
Blue
wavelength (Å)
7,000
Red

For each spectrum, use strips of black paper to represent the spectral lines. Place these strips on the piano keys in the same position as in the spectrum for the element. Now, "play" the chord that results from the markings, and "listen" to the element.

An example of the placement of the lines of hydrogen on a piano keyboard.
The keys marked with "X" are to be played.

Now create chords based on other elements. How do the elements sound differently ?

How is this model different from the true patterns of emission from the elements ?

(Color illustrations for this activity, and a full color spectrum to use on the keyboard, are available at http://imagine.gsfc.nasa.gov/docs/teachers/elements/.)

Cosmic Abundances
(By James Lochner, USRA/NASA/GSFC, Greenbelt, MD)

This activity provides practice for reading and interpreting the log plot of solar system elemental abundances.

A. Refer to the chart of the solar system abundances (Figure 7) on page 10.

From the plot, determine the abundances of the following elements:
1. Hydrogen _____
2. Helium _____
3. Aluminum_____
4. Gold _____

5. How much more hydrogen is there in the solar system than helium?_____
6. How much more hydrogen is there than aluminum? _____
7. How much more hydrogen is there than gold? _____

B. Below is a table of the various processes in the universe that create elements. Also listed are the combined abundances (relative to hydrogen) of that particular groups of elements.

Process	Predominant elements	Combined Abundance
Big Bang	H, He	1.10×10^{12}
Small Stars	C,N,O	1.19×10^{9}
Large Stars	other elements with atomic numbers up through Fe	2.56×10^{8}
Supernovae	elements with atomic numbers above Fe	1.94×10^{6}
Cosmic Rays	Li, Be, B	2.68×10^{3}

What are ways to illustrate or plot these data?

IV. Activity Answers/Assessment Guide

Grandma's Apple Pie

Encourage students to look to use a family recipe for apple pie. Elemental composition of the ingredients can be found on the Web or in reference books.

An apple pie is made primarily from apples, flour, and sugar. Because these are all organic, the primary elements are hydrogen, carbon, nitrogen, and oxygen. Carbon, nitrogen, and oxygen are made in small stars and large stars, while hydrogen is made in the big bang. Students may trace carbon, nitrogen and oxygen back to helium, and then back to hydrogen via the fusion processes in the stellar life cycle.

Another ingredient often used is salt. This introduces sodium and chlorine. These elements are made in the red giant stage of a large star's life. Students may trace these to the helium produced during the first stage of the star's life, and ultimately back to hydrogen.

Students may also find varying amounts of other elements. For example, milk is often brushed on the crust. Hence, there is a trace amount of calcium, which is produced in large stars. In the apples themselves, students may also find calcium, phosphorous (both of which are produced in large stars), and iron (produced in large stars and supernovae).

Suggested grading:
> 40% Accuracy of scientific content
> 20% Quality of presentation
> 20% Creativity of the artistic element
> 20% Successfully addressing all aspects of the problem.

Kinesthetic Big Bang

Answers to the Follow-up Questions

1. The atoms continue to move with kinetic motion.

2. Actually there were trillions of hydrogen and helium atoms formed right after the Big Bang. The small number of students in this model would be a tiny part of the total. Most electrons, positrons, and neutrinos were not included.

3. The universe is not hot and dense enough for further joining of nuclei (fusion) in this way.

4. Inflation fluctuations were frozen into space-time. That means they were converted into slightly denser and slightly less dense regions. The force of gravity locally collected large groups of the atoms to clump together.

Nickel-odeon

The lines in the spectrum of an element represent the energy released when an electron makes a transition between atomic energy levels.

Different elements have different patterns of lines because each element has different atomic energy levels.

The elements can "sound" differently depending on which high or low notes the spectral lines fall.

This model differs from true emission line patterns in at least these ways:
- On a normal piano keyboard, it is difficult to portray the different intensities of the lines.
- In the visual spectrum, blue may be on the left, and red on the right (if plotted by wavelength, where smaller wavelength values are on the left). Higher light frequencies are at the blue end of the spectrum. In this mapping onto the piano keyboard, high frequency light is mapped onto the low frequency sounds.

Cosmic Abundances

A. The abundances can be read directly from Figure 7 (page 10), keeping in mind that the logarithm of the abundances is plotted.

1. If N_H represents the abundance of hydrogen, then the plots shows $\text{Log } N_H = 12$. Inverting the logarithm gives the abundance of hydrogen as $N_H = 10^{12}$.

2. Likewise, $\text{Log } N_{He} = 11$, and $N_{He} = 10^{11}$,

3. $\text{Log } N_{Al} = 6.5$, and $N_{Al} = 3.16 \times 10^6$,

4. $\text{Log } N_{Au} = 1.0$ (estimating from the plot) and $N_{Au} = 10$.

To compare the abundance of an element relative to hydrogen, take the ratio of the abundances:

5. To compare the abundance of helium to that of hydrogen, compute $10^{12}/10^{11} = 10$. So there is 10 times more hydrogen than helium. This can also be done using logarithms,
$$\text{Log } (N_H/N_{He}) = \text{Log } N_H - \text{Log } N_{He} = 12 - 11 = 1.$$
Inverting the result gives $\text{Log } 1 = 10^1 = 10$.

6. Likewise, for hydrogen and aluminum, compute $10^{12}/(3.16 \times 10^6) = 3.16 \times 10^5$, or
$$\text{Log } (N_H/N_{Al}) = \text{Log } N_H - \text{Log } N_{Al} = 12 - 6.5 = 5.5,$$
There is $\text{log } 5.5 = 10^{5.5} = 3.16 \times 10^5$ times as much hydrogen as aluminum.

7. For hydrogen and gold, compute $10^{12}/10 = 1 \times 10^{11}$, or
$$\text{Log } (N_H/N_{Au}) = \text{Log } N_H - \text{Log } N_{Al} = 12 - 1 = 11,$$
There is $\text{log } 11 = 1 \times 10^{11}$ times as much hydrogen as gold.

B. Because of the wide range of values, data like this are often plotted logarithmically. The Solar System abundance bar graph on page 10 is an example of this. These values may likewise be plotted in a bar graph, or as a line graph, with the y-axis for the abundance plotted logarithmically.

Another way to plot it is to add all the logarithmic values (12+9+8+6+3=38), and represent each particular process as a fraction of the total. So H and He from the big bang are 12/38=32 % of the total. A pie chart can be constructed to represent these percentages. This is what's been done along the rim of the circle on the "What is Your Cosmic Connection to the Elements?" poster.

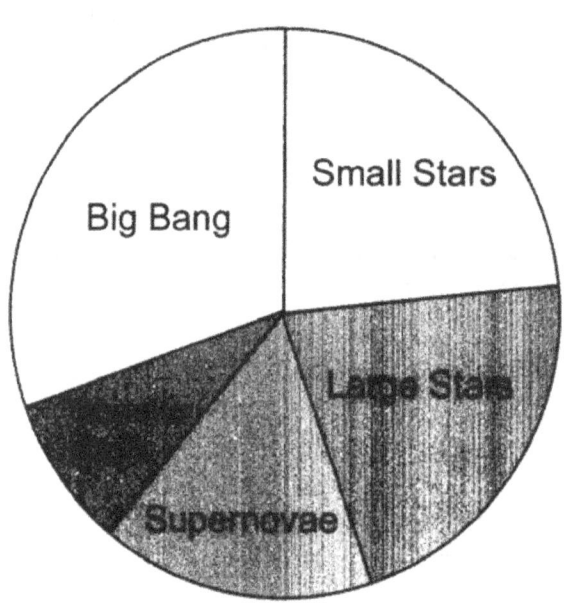

Pie chart representing the "logarithmic percentages" of each process.

V. Glossary

Big Bang – a theory of cosmology in which the expansion of the universe started with a primeval explosion. It is supported by our current understanding of the composition and structure of the universe.

Convection – a process by which warmer, less dense gas rises, and cooler, denser gas falls. This process occurs in the outer layers of some stars.

Cosmic Ray – a nuclear particle or the nucleus of an atom which is traveling through space with very high energies (often close to the speed of light).

Deuterium – an isotope of hydrogen which consists of a proton, neutron and electron. The proton-neutron nucleus is also referred to as deuteron.

Fission – a process in which heavy atomic nuclei break apart to form two or more lighter nuclei. The excess mass of the parent nucleus compared to the resulting nuclei is converted into energy.

Fusion – a process in which atomic nuclei combine to form heavier elements. The excess mass of the colliding nuclei compared to the resulting element is converted into energy.

Heavy Element – to an astronomer, any element other than hydrogen (or possibly helium).

Nucleon – either of the particles found in an atomic nucleus, i.e. a proton or neutron.

Neutron Capture – a process in which a neutron is captured by an atomic nucleus, producing an isotope of the element. If the isotope is unstable, the nucleus will decay into a different element.

Nova – an explosion that occurs on the surface of a white dwarf, due to the accumulation and subsequent fusion of hydrogen on its surface.

Planetary nebula – a shell of gas ejected from stars like our Sun at the end of their lifetime. This gas continues to expand away from the remaining white dwarf.

Radioactive Decay – A process by which an element is converted into a lighter element.

Red giant – a star having a large diameter and relatively cool surface, which results after hydrogen burning has ceased in a star. A red giant has a core in which helium is fusing into carbon.

Red supergiant – a very large star which results after helium fusion has ceased in the core of a very massive star (more than 15 times the mass of our Sun).

Stellar (or solar) wind – the ejection of high-energy particles, including atomic nuclei, from the surface of a star (or the Sun).

Supernova – the explosion of a star. A Type II supernova occurs when nuclear fusion can no longer occur in the core of a massive star, which occurs after the core has fused elements into iron. In a Type I supernova, a white dwarf explodes after accumulating material from a nearby companion star.

White dwarf – the final state of a star about the same mass as our sun after it has exhausted its nuclear fuel of hydrogen and collapsed.

VI. Additional Resources

Web Sites
The chemical elements and the Periodic Table
http://www.wikipedia.org/wiki/Discovery_of_the_chemical_elements - Properties of each of the elements, with information about their discovery.

http://periodictable.com/pages/AAE_History.html - A brief history of the periodic table.

http://www.chemsoc.org/viselements/pages/history.html - An extensive history of the periodic table.

http://genesismission.jpl.nasa.gov/ - The Education area of this web site for the Genesis mission contains a number of classroom materials on the periodic table and cosmic chemistry.

The Big Bang and Early Nucleosynthesis
http://wmap.gsf.nasa.gov/ - The "Universe" section of this site for the Wilkinson Microwave Anisotropy Probe provides information on the Big Bang and nucleosynthesis

Fusion
http://casswww.ucsd.edu/public/tutorial/Nukes.html - An overview of nuclear fusion reactions that occur in stars.

Spectra of Atoms
http://home.achilles.net/~jtalbot/data/elements/ - Emission line spectra of a wide range of atoms.

Life Cycles of Stars
http://imagine.gsfc.nasa.gov/docs/teachers/lifecycles/stars.html

Cosmic Evolution from the Big Bang to Human Kind
http://www.tufts.edu/as/wright_center/cosmic_evolution

Cosmic Rays
http://helios.gsfc.nasa.gov/

Magazines
"We are all Star Stuff", Neil F. Comins, Astronomy Magazine, Jan 2001
Element building in stars.

"Tabling the Matter" by Jim Caddick, The Exploratorium Quarterly, Summer 1992.
Mendeleev and the development of the periodic table.

Video
"The Strange Case of the Cosmic Rays", 1957, directed by William T. Hurtz and Frank Capra, produced by Bell Labs as part of their Science Film series.

VII. Acknowledgements

The poster and booklet were developed by the Education and Public Outreach team at the Lab for High-Energy Astrophysics, NASA/GSFC, under the direction of Dr. James Lochner. Graphic design for the poster was done by Karen Smale. The booklet was assembled by Meredith Bene.

The classroom activities were developed by the attendees of the "Elements 2002 Educator Workshop", August 5-9, 2002 at NASA/Goddard Space Flight Center. This workshop was supported by the U.S. participation in the European Space Agency's INTEGRAL mission. We thank Dr. Ilan Chabay, The New Curiosity Shop, for invaluable assistance with the workshop, and for assistance given to the workshop participants in the development and review of the activities.

Special thanks to our model for the poster, Lai Li, Chesapeake Sr. High School, class of 2002, Pasadena, Maryland, and to our photographer Ms. Geri Cvetic (Chesapeake Sr. High)

www.ingramcontent.com/pod-product-compliance
Lightning Source LLC
Chambersburg PA
CBHW081811170526
45167CB00008B/3396